THE WOODWORM PROBLEM

THE RENTOKIL LIBRARY

The Woodworm Problem
The Dry Rot Problem
The Insect Factor in Wood Decay
The Conservation of Building Timbers
Pests of Stored Products
Household Insect Pests
The Cockroach (Vol. I)
Wood Preservation—A guide to the meaning of terms
Termites—A World Problem
Pest Control in Buildings—A guide to the meaning of terms

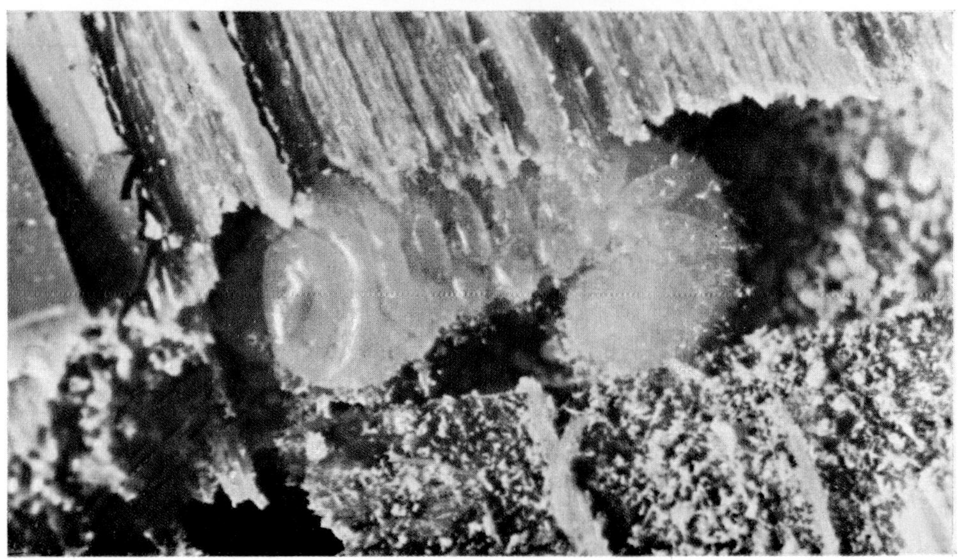

Adult and larva of *Anobium punctatum* much enlarged. Although called the Common Furniture Beetle, this species is by far the most important pest of softwood buildings in Britain